SPOOKY
COUNTING
BOOK
1 to 20

ISBN 978-0-646-89955-8

SPOOKY
COUNTING
BOOK
1 to 20

Frances Mackay

1

ONE scary boy dressed as Frankenstein.
What other scary monsters do you know?

2

TWO young witches.
How many spiders can you count?

3

THREE warty witches.
How many hats are there?

4

FOUR flying witches' cats.
How many broomsticks can you count?

5

FIVE spooky ghosts.
How many look happy?

6

SIX weird monsters.
Which monster has six teeth?

7

SEVEN orange jack-o'-lanterns.
How many green stalks can you count?

8

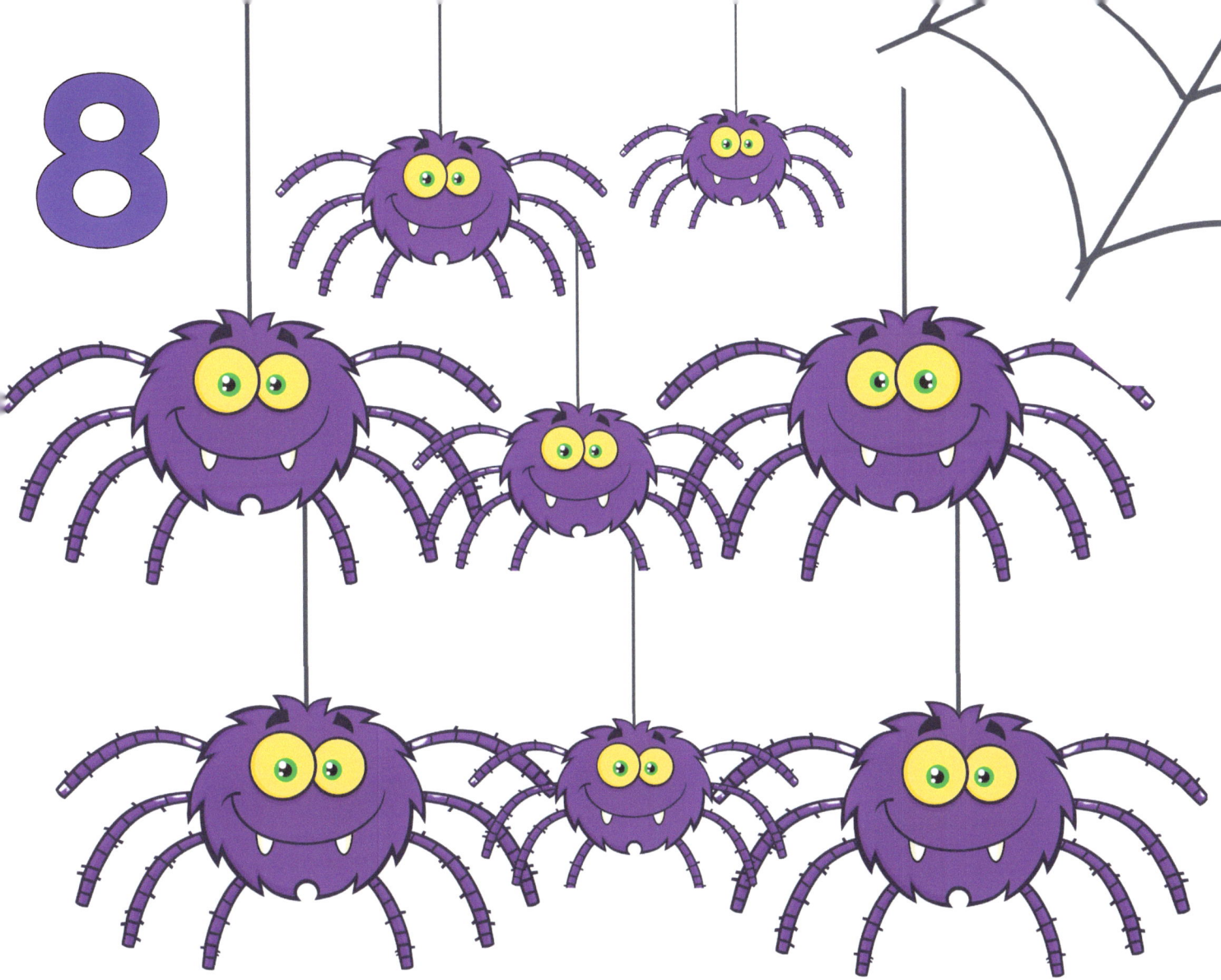

EIGHT purple spiders.
How many legs does each spider have?

9 NINE trick or treat costumes.
How many pumpkins can you count?

10

TEN animals dressed for Halloween.
How many animals can you name?

11

ELEVEN black bats flying around a haunted house.
How many trees can you count?

12

TWELVE dancing skeletons.
How many are upside down?

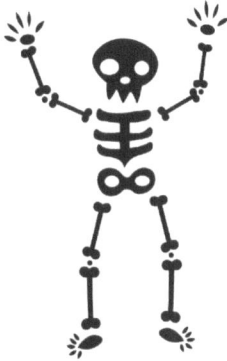

13

THIRTEEN scared cats.
How many are facing RIGHT? ➡

14

FOURTEEN happy ghosts.
How many are facing LEFT? ←

15

A big monster with **FIFTEEN** teeth.
How many warts on her legs?

16

SIXTEEN Halloween balloons.
How many have a skull on them?

17

SEVENTEEN sticky spider webs.
How many spiders can you count?

18

EIGHTEEN spooky cupcakes.
Which one would you eat first?

19

NINETEEN flying witches. How many more are needed to make 20?

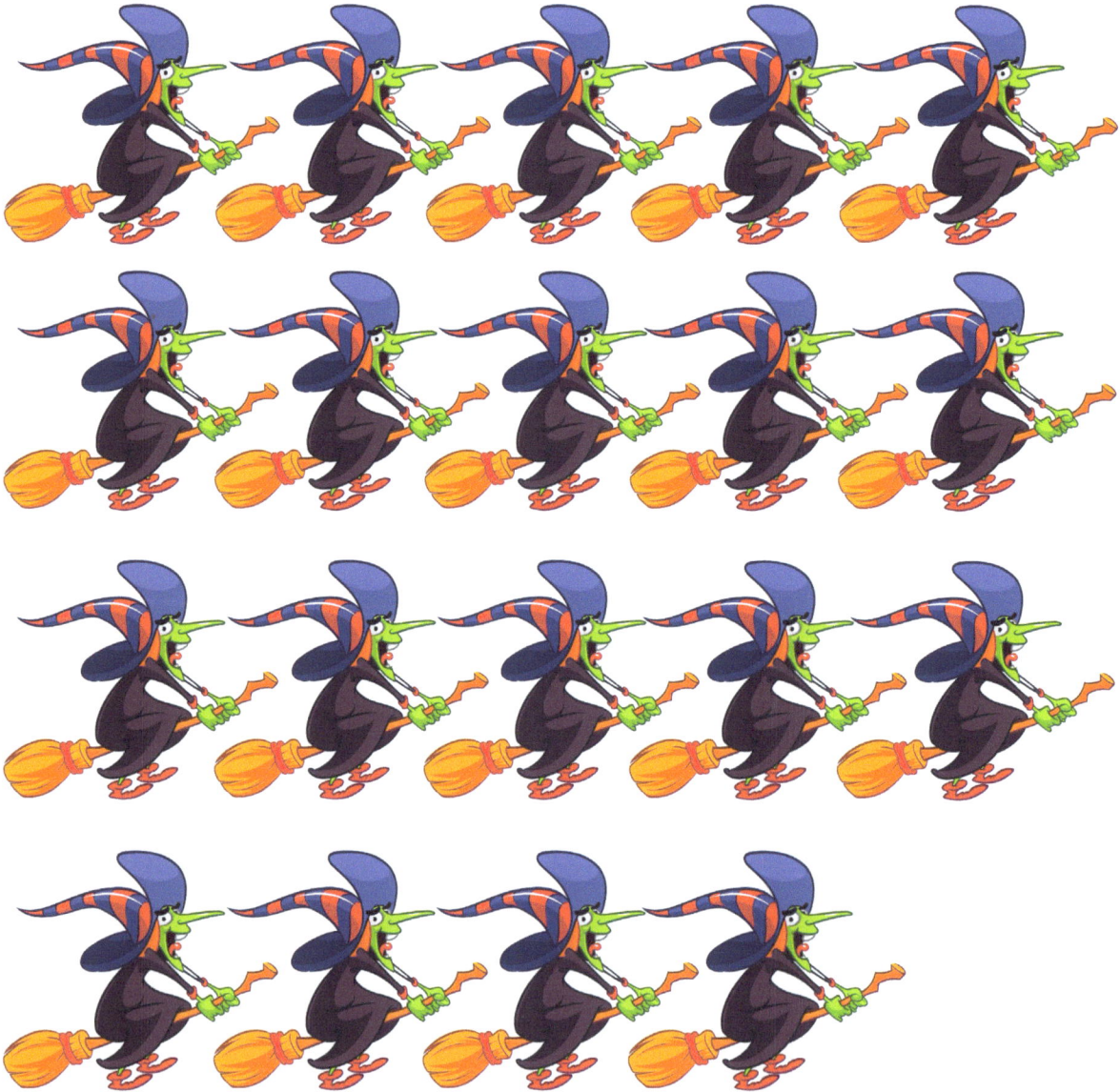

20

TWENTY colourful witches' hats. Which one do you like best?

Now you can count from 1 to 20.

1 2 3 4 5 6 7

8 9 10 11 12

13 14 15 16

17 18 19 20

Well done!

ABOUT THE AUTHOR

Frances Mackay is the author of more than 90 teacher resource books and has written several picture books, activity books and information books.

She was a primary school teacher for 20 years in Australia and the UK. She loves writing books that make learning fun!

facebook:
@francesmackaychildrensauthor
instagram:
@frances.mackay_author

Read more about Frances and grab some FREEBIES at:
www.francesmackay.com

Frances now lives in Tasmania, Australia.

SpooKy

DID YOU ENJOY THIS BOOK?

Your feedback helps me provide the best quality books and helps other readers like you discover great books.

You can leave a review online or send it direct to me at:
frances@francesmackay.com

I read and appreciate each one.

THANK You! ☺

Grab your FREE
10-page
Spooky Colouring Pack
when you visit
www.francesmackay.com

BOOKS BY FRANCES MACKAY

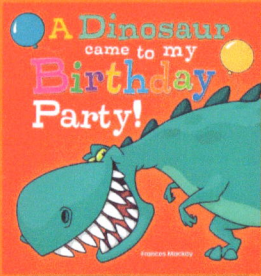

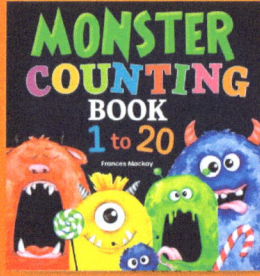

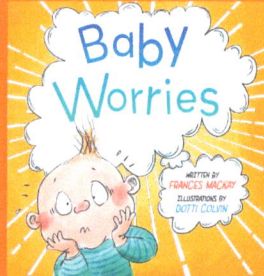

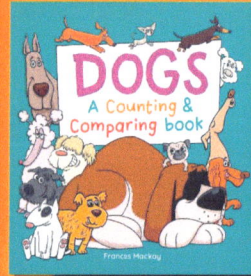

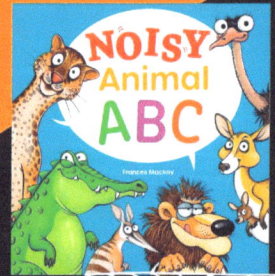

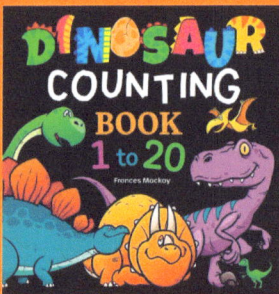

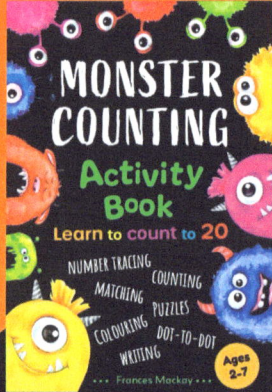

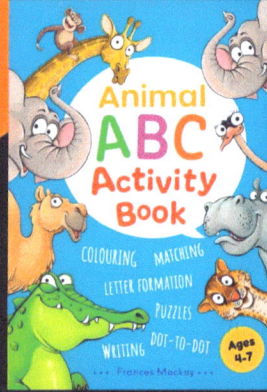

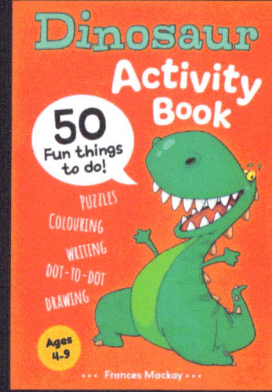

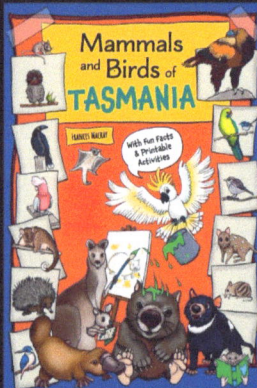

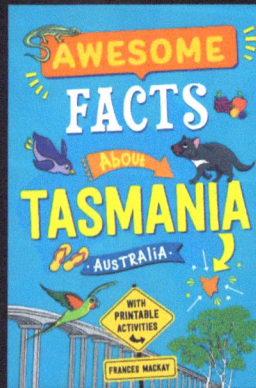

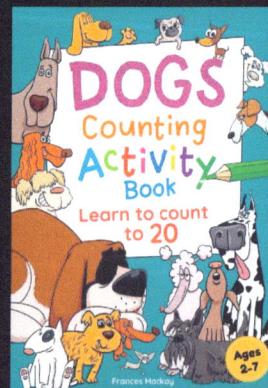

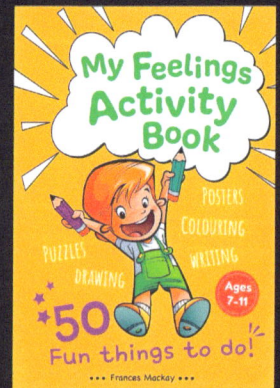

A Dinosaur came to my Birthday Party!

MONSTER COUNTING BOOK 1 to 20
Frances Mackay

Baby Worries
WRITTEN BY FRANCES MACKAY
ILLUSTRATIONS BY DOTTI COLVIN

DOGS
A Counting & Comparing book
Frances Mackay

NOISY Animal ABC
Frances Mackay

DINOSAUR COUNTING BOOK 1 to 20
Frances Mackay

MONSTER COUNTING Activity Book
Learn to count to 20
NUMBER TRACING   COUNTING
MATCHING   PUZZLES
COLOURING   DOT-TO-DOT
WRITING
Ages 2-7
••• Frances Mackay •••

Animal ABC Activity Book
COLOURING   MATCHING
LETTER FORMATION
PUZZLES
WRITING   DOT-TO-DOT
Ages 4-7
••• Frances Mackay •••

Dinosaur Activity Book
50 Fun things to do!
PUZZLES
COLOURING
WRITING
DOT-TO-DOT
DRAWING
Ages 4-9
••• Frances Mackay •••

Mammals and Birds of TASMANIA
FRANCES MACKAY
With Fun Facts & Printable Activities

AWESOME FACTS About TASMANIA AUSTRALIA
WITH PRINTABLE ACTIVITIES
FRANCES MACKAY

DOGS Counting Activity Book
Learn to count to 20
Ages 2-7
Frances Mackay

My Feelings Activity Book
POSTERS
COLOURING
WRITING
PUZZLES
DRAWING
Ages 7-11
50 Fun things to do!
••• Frances Mackay •••

www.ingramcontent.com/pod-product-compliance
Lightning Source LLC
Chambersburg PA
CBHW041557040426
42447CB00002B/204